F8 Crusader in Action

by Lou Drendel

squadron/signal publications

1115 CROWLEY DR., CARROLLTON, TEXAS 75006

ISBN 0-89747-006-0

If you have photographs of the aircraft, armor, soldiers or ships of any nation, particularly wartime snapshots, why not share them with us and help make Squadron/Signal's books all the more interesting and complete in the future. Any photograph sent to us will be copied and the original returned. The donor will be fully credited for any photos used. Please send them to: Squadron/Signal Publications Inc., 1115 Crowley Dr., Carrollton, TX 75006.

Photo Credits:
- U. S. Navy
- U. S. Marine Corps
- Lou Drendel
- Paul Stevens
- Chance-Vought
- L. T. V.
- Arthur Schoeni

Introduction

Chance Vought was not known for its reliance on standard solutions to unique aeronautical problems, (witness the F4U **"Corsair"**, the XF5U **"Flying Pancake"**, and the F7U **"Cutlass"**) and the emergence of the Crusader design preserved Vought's reputation for ingenuity. The Crusader came to life as the winner of an eight company competition for the contract to build the Navy a new interceptor. The Navy had announced the requirement for a new supersonic day interceptor in 1952, and, given the state of the art at that time, could almost have been accused of engaging in flights of fancy for asking what they did of the design. In addition to a level flight top speed in excess of mach 1, the new fighter would have to be tough enough to withstand the rigors of carrier operations, simple enough to be maintained easily and fly slow enough to land aboard a carrier without undue difficulty. As it turned out, the last of these requirements was the toughest to meet, and resulted in the most ingenious design feature of the Crusader. It took the engineers at Vought just five months to come up with the contract-winning design.

The aircraft they proposed was unique in several respects. Foremost among these was the variable incidence wing. The whole wing of the Crusader can be hydraulically raised 7 degrees to achieve a high angle of attack without destroying over the nose visibility for the pilot. In addition to this, there are leading edge slats and drooped ailerons which also provide added low speed lift. The pilot sits well out in front of the wing, so much so in fact, that he cannot see the wing tips easily. This has resulted in several successful wings-folded flights! Titanium and magnesium were used wherever possible to hold down weight and the prototype Crusader carried the unheard-of-for-fighter-planes total of 1,165 gallons of internal fuel.

If the Navy was enthralled with the Crusader design, Chance Vought was positively ecstatic! They were so sure of its success that they broke away from the United Aircraft Group to become independent and stake their whole future on the Crusader. History has validated their decision. The first prototype flew in 1955, acheiving supersonic speeds on its first flight. It was an auspicious beginning for the test program, which ran its course without serious setback. By March of 1957, F8-As were rolling off the production line at the rate of eight per month. The Crusader had begun a career which, at its zenith, would see the Chance Vought fighter equipping 36 Navy and Marine Corps squadrons.

F8U-1, with Cdr. R. W. Windsor at the controls, during carrier qualifications aboard **USS Forrestal**, in April of 1956. Cdr. Windsor also won the Thompson Trophy in 1956, setting a new national speed record of 1,015 mph in the Crusader. (Chance Vought)

One of two Crusader prototypes. The drooped ailerons and leading edge slats are clearly visible. (Chance Vought)

The unsuccessful half of **Project Bullet.** F-8A flown by Lt. Charles Demmler was unable to complete the coast to coast speed run, due to damage incurred during in-flight refueling. The other Crusader, an RF-8A flown by John Glenn, completed the supersonic crossing of the United States. (Paul Stevens)

The Crusader chalked up a notable first as Cmdr. George Talley, **CAG Air Group One**, made the first landing on the nuclear carrier **USS Enterprise.** (US Navy)

YF-8A at **NAS Johnsonville, Pa.** in 1965. (Paul Stevens)

Crusaders of **VF-62**, the first aircraft to launch and recover on the new **"Big E"**. (Chance Vought)

F-8D incorporated numerous changes from the "C" model, one of which was the "Y" rack that enabled the Crusader to carry four **Sidewinder** AAMs, instead of the original complement of two. (Chance Vought)

The F-8D also had improved avionics, including a push-button autopilot. The **"Mighty Mouse"** rocket pack was deleted in favor of increased fuel capacity. (Chance Vought)

One of 130 F-8Bs to be built. The "B" model had improved electronics, which gave it a limited all-weather capability. (LTV)

F8U-1 of **VF-103**, at **Langley AFB** in 1961. (Paul Stevens)

Head-on view of the F-8E. The "E" incorporated numerous modifications, including detachable wing pylons and additional avionics, which allowed the Crusader to employ the **AGM-12 Bullpup** missiles shown in this view. Note the geometry of the "Y" racks, in this case mounting **Zuni** air-to-ground missiles. The nose mounted boom was used for test programs only.(LTV)

A telephoto lens creates an impressive if somewhat distorted portrait of this F-8D about to launch from **USS Constellation**. The Crusader is from **VF-124**, a west coast training squadron. (LTV)

An F-8A of **VF-32** aboard the **Saratoga** for an Atlantic cruise in the early sixties. Early Crusaders were equipped with 2.75 in. **"Mighty Mouse"** rocket packs, mounted in the fuselage speed brake. (Chance Vought)

F-8E at **Patuxent River**, 1969. The small window on the underside of the nose cone is the gun camera aperture. (Lou Drendel)

Launch sequence. At left F-8s of **VF-174** run through various operations leading to catapult launch. Catapult crew attaches launch bridle and hold-back cable. (top) Final check of cables is made while pilot completes his check list. (middle) Crusader roars down the flight deck under the combined thrust of **J-57** engine and steam catapult. (Harvey Swedowski via Paul Stevens)

Vought's entry in the mach 2 interceptor competition was the Crusader III. Although it bore a superficial resemblance to the operational Crusader, the F8U-3 Crusader III was an all new airplane. It lost out in the competition to the more powerful and sophisticated McDonnell Phantom II. (Chance Vought)

A **VF-32** Crusader about to catch a wire aboard the **Saratoga.** The lights of the mirror landing system are visible in the foreground. (LTV)

This Crusader of **VF-124** has just recovered, and is being directed to a parking spot as a Phantom turns onto a final approach. (Harvey Swedowski via Paul Stevens)

F-8E undergoing carrier suitability tests by the **Naval Air Test Center.** The F-8E was the final growth version of the design and numerous improvements gave the Crusader all-weather intercept and ground attack capability. Note the **"Mad Bomber"** motif on the fuselage side. (LTV)

This underside view of an F-8E shows the position of the fully retractable tail hook (LTV)

An F-8D about to touch down. The F-8D was the first Crusader to mount the infra-red scanner above the nose cone. It also carried the then-new approach power compensator. (LTV)

A test version of the F-8E at **Patuxent River** demonstrates the ground attack capabilities of the Crusader, mounting **Snakeye** bombs on triple ejector racks. The F-8E is able to carry up to 4,000 pounds of bombs on wing hardpoints. (US Navy)

Three views of the sole example of the TF-8A Crusader dual control trainer. The TF-8A is a remanufactured F-8A. It is now in permanent residence at the **Navy Test Pilot School** at **Patuxent River.** (LTV and David Ostrowski)

The pilot of this **VF-124** F-8E and the flight deck crew of **USS Constellation** take a momentary break as **Connie** completes her turn into the wind. When the Captain of the carrier has the wind on her nose launch operations will continue. (LTV)

An F-8J of **VX-4** refueling from an EKA-3B. The F-8J is a remanufactured F-8E. (via George Letztzer)

An F-8E(FN) aboard **USS Shangri-La** for suitability tests. The French bought 42 F-8Es for use on carriers **Clemenceau** and **Foch.** French Crusaders incorporated boundry layer air and double droop wing leading edges for even better low speed handling. (LTV)

An immaculate F-8D, one of 152 "D" models manufactured. A total of 89 were remanufactured and redesignated F-8H. (US Navy)

An F-8H of a Naval Reserve squadron swoops onto the deck of the **USS John F. Kennedy**. The remanufacture program for Crusaders guarantees the **LTV** fighter operational life well into the seventies. Note the deployed emergency generator in this view. (under fuselage pylon) (US Navy)

An F-8L (remanufactured F-8B) taxies out for a flight test at **LTV's** Dallas plant. 63 F-8Bs received a new lease on life through the remanufacture program. (LTV)

An F-8D undergoing flight tests. The scoops atop the tailpipe are for afterburner cooling. (US Navy)

The F-8C also benefitted from the remanufacture program, 82 of the older models becoming F-8Ks (LTV)

The Crusader in Combat

When the Crusader III lost out to the McDonnall Phantom II, thereby assuring the ascendancy of the F-4, a rivalry between the single engine, single pilot Vought fighter and the super-sophisticated Phantom was born. Crusader pilots ridiculed the big McDonnell fighter, citing the twin engine, two pilot, no gun layout as proof positive that the Phantom was not a fighter plane. A fighter-bomber maybe, but a real interceptor . . . never! The Vietnam War provided an arena for the pilots of both fighters to test the mettle of their mounts.

Crusaders were among the first aircraft to launch retaliatory attacks against North Vietnamese torpedo boats after the now famous **Tonkin Gulf Incident.** During the Vietnam War Crusaders continued to serve with distinction on the smaller, **Essex** class carriers, whose flight decks were unable to withstand the punishment of repeated Phantom landings.

One of these carriers was the **USS Oriskany**, which carried two F-8 squadrons in her air wing. The Commanding Cfficer of one of these Crusader squadrons, **VF-162**, was Captain (then Commander) Richard Bellinger. Dick Bellinger was a seasoned veteran of many combat missions before the 1966 **Oriskany** deployment to **Westpac.**

At the outbreak of World War II he was a student at **Tufts College**, Medford, Massachussets. Like so many other young men at the time, he heeded his country's call to arms, enlisting in the Army Air Corps. He eventually became a bomber pilot, flying B-17s and B-25s. After the War he went back to **Tufts** to complete his education. In addition to his more scholarly activities at **Tufts**, he became a member of the **Naval ROTC** unit. He graduated from **Tufts** in 1948 with a BA in psychology and a commission in the **Naval Reserve**. He requested, and received assignment to Navy pilot training, eventually earning his wings of gold in 1950.

Two deployments to the Med followed, as a fighter pilot, flying F2H Banshees. Eight deployments to **Westpac** followed, the first being a combat deployment in the Korean War. Bellinger flew 77 combat missions in Korea, including the strike on the bridges at **Toko-Ri**, made famous by James Michner's book of the same name. From 1953 to 1955 he was an instructor in the **NROTC** at **Harvard.** The period from 1955 to 1957 saw him assigned to the **USS Badoeng Strait (CVE-116)**, as CIC/Air Intelligence Officer, taking part in the **Bikini Atomic Test Deployment.** In 1957 he joined **VF-142**, one of the first Crusader squadrons. The **Westpac** deployment that followed began a long and eventful association with the Vought fighter.

USS Oriskany steaming in the **Gulf of Tonkin** during the bombing campaign against North Vietnam in the mid-sixties. (US Navy)

In 1959 Bellinger began a long period of scholastic activity that included completion of **General Line School**, studies at the **Naval War College**, Newport, R.I., (where he also instructed) and culminating in the award of a Masters Degree from **Boston University** in Political Science.

In 1963 he was back on the line, as Operations Officer for **Carrier Air Wing Eleven**, participating in pre-Vietnam War operations. (Which, if the truth is known, probably included a goodly number of combat missions.)

At the time of the official U. S. entry into Vietnam combat, he was Executive Officer of **VF-162**, aboard the **Oriskany**. He would eventually become CO of **VF-162** and, following the untimely demise of the air group commander in the **Oriskany** fire, CAG **Air Group Sixteen**. During two **Westpac** deployments aboard **Oriskany**, Bellinger was shot down by a MiG-17, and evened the score by becoming the first Crusader pilot to down a MiG-21.

By the summer of 1966 the bombing campaign against North Vietnam had assumed major proportions. The North Vietnamese defenses were improving daily, and their MiG fighter forces were taking an increasingly heavy toll of U. S. bombers. It had become customary for the carrier air wings to conduct three **Alpha** (major) strikes a day against selected targets. The MiGs became most aggressive on the last strike of the day, since it was conducted close to sundown when visibility was poorest. The MiGs, with their Ground Controlled Intercept procedures, have a definite advantage under these conditions. The bomb laden attack aircraft were often unable to jettison their bombs and defend themselves before being attacked by the MiGs, and their fighter escorts often found themselves in the wrong place at the wrong time. Commander Bellinger suggested that the attacking forces use the MiG tactics to turn the tables on the enemy. He proposed that the fighters lead the attack force into the target area, with the bombers well to the rear. The enemy radar operators, unable to distinguish between the fighters and bombers, would more than likely vector the MiGs onto the fighters, hoping to blunt the American attack. His plan was approved and on 17 July 1966, he got a chance to test it in combat.

Bellinger led a three plane element on the last strike of the day. (The fourth plane didn't get off the deck.) As the three Crusaders approached the target in a loose "V" formation, they were jumped by five MiG-17s.

There was a solid overcast at 2500 feet and the light was failing fast as the combat began. The Crusaders had been flying at 2000 feet and the GCI controller had vectored the MiGs onto the Crusader's tails before having them descend through the clouds. The enemy interceptors broke through the clouds in a sweeping right hand turn that would put them in perfect firing position on the American fighters' tails. Fortunately, Bellinger had been looking at the right piece of sky at the right time, and spotted them almost immediately. He called for a hard right break into the MiGs. The number two Crusader, flying on Bellinger's left wing, had just lost his radio and didn't hear the call and,

F-8Es of **VF-162** over the California desert on a training mission. (US Navy)

unfortunately, didn't spot the MiGs. As Bellinger and number three broke right, number three automatically assumed the lead. The first two MiGs went after number two, who still hadn't spotted them. Number three, pulling eight G's, cranked his F-8 into position behind the first two MiGs. The third MiG got on number three's tail. The fourth and fifth MiGs maneuvered for position, hoping to get a clear shot at one of the American fighters. By this time number two had spotted the enemy fighters and was busily trying to get them off his tail. The fight had degenerated into a wild free-for-all, with six of the fighters right down on the deck, dodging trees and farm buildings as they attempted to gain the advantage. Speeds remained in the neighborhood of 350 knots and G loads varied from 5 to 8. Number three finally got into position for a shot at the two MiGs on number two's tail. His guns had jammed, probably due to the high G Load, so he fired his **Sidewinders**. One of the disadvantages of the **Sidewinder** is that it is a heat seeker and the enemy tailpipe must be outlined against a clear piece of sky. In this case, there was just too much ground in-

An F-8E of **VF-162**. Marking details are: black rudder with yellow stars, yellow stripe outlined in black at top of vertical stabilizer, black bands with yellow stars on wing tips and under nose. (Neal Schneider via R. M. Hill)

terference, and the missiles went ballistic into the North Vietnamese country side.

Bellinger, with a full load of missiles and cannons that hadn't been adversely affected by the violent maneuvers, radioed that he intended to "Yo-Yo" up over the top, and slide into number three's position for a crack at the two leading MiGs. He pulled the nose of his Crusader up to 2000 feet, cut across the circle, and had just begun to dive onto the MiG's tails when all hell broke loose. The two trailing MiGs, momentarily forgotten in the heat of battle, had gotten on his tail and were blasting the F-8 to pieces! The starboard outer wing panel was literally in shreds, the starboard stabilator had been shot completely off, hydraulic pressure was failing, and the Crusader was stream- a trail of fuel from the seived wing tanks. The MiGs held their fire as the F-8 went into an apparently uncontrollable series of rolls. Bellinger fought for control and, after reducing power and speed to about 225 knots, finally brought the crippled fighter under control. Without another look around, he pulled the nose up into the clouds and set sail for the **Gulf of Tonkin**. The MiGs had lost their quarry, but Bellinger might still wind up in the **"Hanoi Hilton"** if he didn't nurse the mangled F-8 to an area where he had a better chance of rescue. He harbored no illusions about landing the crippled fighter on the carrier. It just couldn't be done. He would have to head for **Danang** and, depending on how the plane handled, either try a landing on their long runway, or eject off the coast. His luck held and he was able to get out over the sea, where he immediately set course for **Danang**, calling "mayday" all the way. Because of his hydraulic problems, he was unable to deploy his in-flight refueling probe, and ran out of fuel about 40 miles short of **Danang**. The Martin-Baker ejection seat worked as advertised, and Bellinger was soon afloat in the **Gulf of Tonkin**. Rescue forces had been alerted and had followed his progress south and, after about twenty minutes in the water, he was picked up by an Air Force helo and flown to **Danang**. He was back aboard the **Oriskany** the following day, certified fit for further combat. His plan to decoy the MiGs into attacking the **MIGCAP** fighters, instead of the bomb-laden aircraft, had worked . . . almost too well for his own good!

Dick Bellinger firing **Zuni** air-to-ground rockets at a Viet Cong stronghold in the Delta of South Vietnam. This action occurred during a tune-up cruise on **Dixie Station** in 1966. (US Navy)

In October of 1966 the **Oriskany** was again on the line, flying daily missions against North Vietnam from **"Yankee Station"** in the **Gulf of Tonkin.** On the 9th of October, the **Oriskany** was asked to provide fighter cover for a strike force from **Intrepid.** Bellinger led the fighter forces, and since the target was relatively close to the sea, he requested an E-1B airborne early warning aircraft be stationed off the coast in position to warn them of the approach of MiGs. To further aid the fighters in closing with the enemy fighters, the E-1 was instructed to report the MiG's position in relation to the target, rather than in relation to **Hanoi** or **Haiphong**, as was the normal procedure.

As the attack aircraft were plastering the target, the fighters were "hiding" down low, under North Vietnamese radar coverage, in the nearby mountains. The ruse worked, as the enemy radar controller vectored his fighter force toward the attackers. MiGs were first reported on the 330 degree radial, 90 miles short of the target. The E-1 kept up a running report of their progress as they approached the target area. As they passed the 40 mile mark inbound, Bellinger led his fighters out to intercept them.

As Bellinger emerged from the mountains, 300 feet above the rice paddies, he spotted the MiGs about 3000 feet above. He pushed the throttle through the detent into afterburner and pulled the nose up, climbing into firing position behind the nearest of the MiGs. Coordinating stick and throttle perfectly, he slid into position behind the enemy fighter, and with a zero deflection shot at the MiG, prepared to blast him from the sky. The MIG leader's wing man called out the danger just then, but instead of breaking left or right, the MiG rolled over and dove for the ground. If he could split-S from this altitude, he would be home free, as the Crusader couldn't possibly hope to match his maneuver. Bellinger matched his wing position, though, and as the Crusader's nose passed throught the horizon, he fired off two **Sidewinders.** Both of the missiles guided perfectly, the first blowing off the MiG-21's wing, the second exploding nearby. But there was no time to watch his vanquished enemy spiral into flaming oblivion, for he was going straight down at a high rate of knots, and the ground was almost too close already. He knew he couldn't pull through the split-S, for even if he was able to withstand the G loads this would impose, he was sure the Crusader would shed its wings. Bellinger snapped the stick over, rolling 180 degrees, and pulled hard. Every spar and joint in the Crusader seemed to groan with the effort, but it pulled out, streaming condensation trails from the wingtips, just over the treetops. With a last look over his shoulder at the burning remains of his victim, Bellinger climbed back to altitude. The remaining MiGs had fled though, and the American fighters, mindful of their responsibility to protect the bombers, did not pursue them.

Bellinger put his combat experience and initiative to good use in another area of operations while on this cruise. The North Vietnamese, with seemingly endless supplies of **SAM** missiles, would often fire salvoes of the radar guided missiles at the attack airplanes. The bomb laden attackers, their maneuverability greatly impaired, often were unable to avoid all the enemy missiles without jettisoning their bomb loads. Bellinger proposed that the fighters accelerate ahead when five minutes short of the target, then "pop up" over the target, simulating the start of a bomb run, The Crusaders, with their greater maneuverability and not required to follow through on a gauntlet-running bomb run, should be able to draw the bulk of the enemy fire and, hopefully, avoid all of the **SAMs.** This plan was put to the test and Bellinger and his wingman were exposed to eight **SAMs** almost immediately. They successfully avoided all of them and the following attack aircraft hit the target without having a single **SAM** fired at them. The Crusader, in the hands of an able and aggressive fighter leader, proved that it was an able adversary for MiGs and **SAMs** both.

Ordnance specialists are loading **Sidewinder** missiles on an F-8E of **VF-52**, aboard **USS Ticonderoga.** The Crusader is also armed with four 20 mm cannon with 144 rounds per gun. **Sidewinders** have proven to be the more effective air-to-air weapon, as the guns tend to jam under the high G loads imposed during dog fighting. Crusader pilots are still fond of referring to their mount as "The last of the gunfighters." (US Navy)

The completion of the **Sidewinder** loading sequence. Note the movable control surfaces of the **Sidewinders.** (US Navy)

A **VF-51** Crusader over the **Gulf of Tonkin.** The depth of the refueling probe bulge on the port side is evident in this photo. (US Navy)

The business end of an F-8E loaded for the air superiority role. (US Navy)

An F-8D of **VF-111**, off **USS Midway**, firing a **Zuni** rocket at enemy positions early in the war. (US Navy)

A colorful F-8E. Tail colors are, from top to bottom: light blue, lemon yellow, green, orange, red. The lightning bolt is red. The aircraft belongs to CAG, **CVW-19.** (Kenji Kubota via George Letzter)

An F-8J of **VF-194** launches from **USS Oriskany**, during 1969 air operations off North Vietnam. (US Navy)

An F-8E of **VF-194** taxies to the active runway at **Danang.** (Neal Schneider via R. M. Hill)

An F-8E of **VF-53.** Tail colors are, front to rear: Red, light blue, lemon yellow, green, scarlet. The aircraft belongs to CAG, **CVW-5.** (Yatsuhashi via George Letzter)

An F-8J of **VF-53** in the landing pattern over **"Bonnie Dick"** during operations in the **Gulf of Tonkin**, circa 1970. (US Navy)

An F-8J aboard **Oriskany**, during a 1971 call in Japanese port. (via George Letzter)

The moment of truth during the recovery of an F-8 aboard **Bon Homme Richard.** The Crusader, streaming condensation trails from its wingtips, swoops over the ramp as the LSO looks on. The advantage of the variable incidence wing is evident here, as the angle of attack necessary to maintain approach speed would severely limit pilot visibility without this unique design feature. (US Navy)

A Marine Corps KC-130 extending the range of a pair of Marine Crusaders. Maximum unrefueled range of the F-8 is 1,400 miles. (USMC)

VMF-235 was one of the first Marine Corps fighter squadrons to operate the Crusader. The F8U-1 is seen here during carrier quals. (via Paul Stevens)

An F-8E of **VMF(AW)-235** on approach for landing. Markings colors as follows: Red nose flash, tail stripe, and ventral fins with white stars. (via George Letzter)

F-8E of **VMF(AW)-235** ready for action at **Danang**, 1967. (USMC)

(Right) A Crusader of **VMF(AW)-235** on the flight line at **Danang** as the pilot and crew chief go through pre-flight procedures. The F-8E is armed with **Zuni** air-to-ground rockets and 2,000 lb. bombs. (USMC)

An F-8E of **VMF-122**. Colors are: Blue rudder and circle, white shield and sword, red cross, black fin tip. This is an early F-8E, sans avionics later added to enable Crusaders to carry **Bullpup** missiles. (USMC)

An F-8E taxies to the runway prior to a mission against enemy units in the **I Corps** area of South Vietnam. A testament to the Crusader's all-weather ability is evident in this picture as the hill tops in the distance are obscured by low hanging clouds. Marine Crusaders played an important role in the around the clock bombing campaign that eventually led to lifting the siege of the Marine garrison at **Khe Sahn**. (via George Letzter)

(Left) A Marine F8U-2 at an air show in 1960. (via Paul Stevens.)

Ten thousand pounds of **J-57** thrust plus the steam catapult of **USS Forrestal** sling an F8U-2 into the air. Though Marine fighter squadrons did not (until recently) operate regularly from carriers, all Marine pilots are required to maintain carrier qualification. (LTV)

F-8Es of **VMF(AW)-312** pose against the backdrop of the San Bernardino Mountains. Marking details are: Dark brown fin tip above yellow and red stripes, black and white checkerboard with red outline, black ventral fins with yellow and red border. (USMC)

An RF-8A of **VFP-62.** Photo Crusaders of **VFP-62** monitored the buildup of Soviet supplied missiles in Cuba during the 1962 crisis. RF-8As are operated by both U. S. Navy Light Photographic Squadrons, which deploy detachments to various Carrier Air Wings. This results in aircraft of the same squadron sometimes being adorned with a variety of markings. Major structural differences in reconnaisance Crusaders include; elimination of cannon and fire control equipment which in turn results in increased cross sectional area. This necessitated the area-ruling of the fuselage, thus producing a "hump-backed" appearance. Photo Crusaders carry 5 cameras plus photo-flash bombs for night time photography. 144 RF-8A Crusaders were built. (via Paul Stevens)

A Marine RF-8A during carrier qualifications. Marine Corps Composite Reconnaisance Squadrons (**VMCJ**) **1, 2, & 3** operated the RF-8. (via Paul Stevens)

A Marine RF-8A remanufactured to RF-8G standard. 53 RF-8As were remanufactured to "G" standards. Marine reconnaisance squadrons eventually transitioned to RF-4Bs. (via Paul Stevens)

An RF-8A of **VFP-63, Det Alpha** approaches **USS Midway** for a landing during 1968 Vietnam combat deployment. (US Navy)

A carrier cycle sequence. RF-8As of **VFP-62** positioned for attachment to the catapult of **USS Forrestal**, (above) about to pounce on the deck, (right) being hauled to a stop as a deck crewman positions himself to disengage the arresting gear, (below) and being directed out of the landing area (below right) (US Navy and Harvey Swedowski via Paul Stevens)

An RF-8G of **VFP-62** roars off the angled deck of a CVA after a "bolter". The tailhook failed to catch a wire and the pilot of the Crusader was forced to go around for another try. Nonchalant attitude of the deck crewmen indicates that this was not a rarity. (Arthur L. Schoeni)

An RF-8G from **USS F. D. Roosevelt** during a 1969 cruise to the Caribbean. The squared off cross section housing the cameras is most evident in this photo. (US Navy)

A **VFP-62** RF-8A during a flyby of the **USS F. D. Roosevelt** in the Med in 1965. (US Navy)

An RF-8G of **VFP-63**. Blue bands, white stars on tail and wingtips. Navy light reconnaisance squadrons will continue to operate the RF-8 well into the 1970s. (via George Letzter)

Reserve Crusaders. An RF-8G of **VFP-306**, one of two Naval Reserve light photographic squadrons. **VFP-306** is homebased at **NARTU Washington, D. C.** (via George Letzter)

A well-worn F-8A of the Naval Air Reserve. Crusaders have been operated by several Reserve units. They will be replaced, at the end of their fatigue endurance life, by F-4s. (via Paul Stevens)

RF-8Gs of **VFP-206** and **306** in the refueling pits at **NAS Cecil Field, Florida** during a 1970 deployment of the two Reserve units. Extensive training at **Cecil Field** was followed by carrier quals aboard **USS F. D. Roosevelt.** (R. Hill)

Epilogue

Despite a radical change in the Navy interceptor concept early in its operational life, the Crusader design was so successful that it was given a longer lease on service life. It proved time and time again that it was a competent clear air mass fighter and an adequate all-weather fighter. Vietnam War combat experience caused Navy pilots to rate the F-8 wing as the best dogfight wing on any United States fighter. A total of 1,261 F-8s were manufactured before the assembly line was closed in 1965. Of these, 448 received a reprieve from retirement through the remanufacture program that saw the installation of newer, stronger wings and landing gear, as well as more sophisticated avionics.

Twenty years after the Crusader design was first proposed, F-8S are still performing and being proposed for new and challenging assignments. In 1970 **LTV** put together a proposal for the Air Force's International Fighter Competition, which in effect was a much modified F-8D. They called it the V-1000 and it lost out to the **Northrop** F-5E.

The Crusader has been employed by **NASA** in the testing of its super-critical wing. In these tests, the variable incidence wing of the original design was replaced by the super-critical wing, which was fixed firmly in place. So far the tests have been successful, thanks in large part to the soundness of the basic Crusader design. If the super-critical wing fulfills its promise of greater speed with less drag, the Crusader will have played a major role in the proving of one of most significant aerodynamic theories of our day.

Crusaders still equip several operational Navy fighter squadrons and some Navy Reserve units. It seems a safe bet that the **Vought** fighter will remain a familiar sight aboard aircraft carriers and at Naval Air Stations throughout the world for years to come.

Several Crusaders were modified for special duties, such as target tugs. F-8A, above, in the eye-catching colors of **VC-7**, (dark blue-gray fuselage, orange vertical fin, wings and stabilators) and F-8L of **VC-7**, below, are two examples. (via George Letztcr)

Whether your interest is World War II or Modern . . . . Aircraft, Warships, Armor or Uniforms . . . the "in Action" series has the books to satisfy your needs . . . and the list is growing, with new titles being added regularly.

More from

squadron/signal publications